Cover Pic: Adapted from an original painting by Ian Kennedy

Published By

David Bradley Management

PO Box 7271

Perth PH2 7YW

UK

1996

A History of the Tay Ferries 1713 - 1966

By

David Sinclair

Edited By

Neil E. Sinclair BSc

Acknowledgements

D.C. Thomson & Co. Ltd.
Dundee City Archives
Bradley Borland
Maisie Mands
Leonora Brown
Gavin Johnston

Foreword

This book plots the course of the Dundee ferry operations in Scotland's East coast estuary of the River Tay. Special attention is given to the mechanised period, from 1820 until the cessation of the service in August 1966, when the Tay Road Bridge was opened by H.M. The Queen Mother, and the ferryboats were sold. In all, thirteen boats were built during this period for the service between Newport and Dundee.

The ferry boats, affectionately known as the 'Fifies', were an essential link, especially in the post war (World War II) years with many people commuting to and from North Fife, which was a popular residential area for Dundee workers.

While conducting the research for this book, many illustrations and photographs were handed over from various collections. Due to the passage of time, and despite strenuous efforts to find and acknowledge the original artist or photographer, it cannot be certain that we succeeded in every case.

Tay Ferry Crossings

Since the times of the early settlements on the banks of the Tay Estuary, there had always been boats crossing between North Fife and Dundee. These early boats were privately owned craft of all shapes and sizes.

Various crossing points were used over the years, with Dundee, Broughty Ferry and Errol on the North shore and Ferry-Port-on-Craig (Tayport), Sea Myles (Newport), Woodhaven and Balmerino on the Fife side of the estuary, Woodhaven being the most popular passage to Dundee in the early days.

To the considerable annoyance of the Dundee Town Council, the Ferry-Port-on-Craig (which derived its name from the rock on which the ferry landing was erected) to Broughty Ferry route proved very popular with the coal miners in Fife due to cut price fares being offered. Cattlemen driving large herds to the South for slaughter, also found this passage the most convenient due to ample grazing being available on both sides of the river.

Organised Sailings

By the beginning of the eighteenth century, in the year 1713, the need for a more direct communication between Dundee and Edinburgh, by the way of Fife, was felt necessary. A suggestion by the Dundee Town Council was taken up by the Guildry of Dundee, who agreed to establish a more regular ferry service than the one existing at the time. This new service would run from Dundee to Sea Myles, which lay between the new pier and the old harbour of Newport-on-Tay. Ground was purchased on which piers and a house were built, the whole project taking about eighteen months to complete.

This landing at Newport was, for sometime, to prove to be under used, with the other passage from Dundee to Fife, the Woodhaven Ferry, proving much more popular. This was mainly due to the existence of a good road between Woodhaven and Cupar.

In 1787, a turnpike road was constructed to connect Newport with the Cupar road at St. Michaels Wood, a distance of about 4 miles from the Newport pier. This changed the fortunes of the Newport crossing, with it now being considered the main link for passengers journeying from the South to Dundee and the Northeast of Scotland.

Changes

By 1815, twenty-five boats were plying the two routes, from Dundee to the river Tay's south shore at Newport and Woodhaven, giving employment to some one hundred men and boys. The boats were generally owned by their crews, who shared in the profits derived from their labours.

There were no fixed sailing times, with the boats awaiting a full compliment on board before departing. The passenger fare of 9d (about 4 pence) was, in the face of the fierce competition, often reduced to 6d, in order to fill the boat and allow it to sail.

The small passenger-only boats were pinnaces, fitted with masts and carrying lug sails similar to fishing boats. The larger boats were yawls - strongly built, sloop rigged, with an ample spread of canvas, and were adapted for the transportation of horses, cattle, carts and carriages.

At this time, the two landings on the south shore of the estuary were known as East Water (Newport) and West Water (Woodhaven) Ferries. With favourable conditions the crossings were, at best, far from comfortable and, indeed, were often fraught with danger in the open estuary and perilous currents present on the river Tay.

Early Pinnace passenger boat

Safety and Organisation

It was on the morning of Sunday 28th May 1815, that the attention of those interested in the safe passage of travellers using the ferry crossings was moved to action. A pinnace, commanded by John Spalding, capsized when about half a mile from the Fife shore. Of the twenty-five people on board, only seven survived. The convenor of the Joint Committee of Enquiry, Mr. William Berry of Tayfield (a wealthy Dundee businessman) investigated the entire business of safety and operation of the ferry services. He reported that during the year of 1816-17, a revenue of £3,177 was derived from the passage of 91,663 passengers, 13,612 cart loads of goods, 2,659 horses, 2,321 head of cattle, 4,257 sheep, 126 pigs and 670 coaches and gigs.

The report concluded that sufficient revenue was derived to allow a more efficient service to be operated, and to provide better landings on both sides of the river. Mr. Berry also noted that with the licences easily obtained, there was no shortage of boats, now numbering twenty-eight, of all shapes and sizes.

The report was submitted to the Justices of the Peace for Fife and Forfar (of which Dundee was a part) at a meeting held in Newport, on 13th September 1817.

Provost Riddoch of Dundee, acting as chairman, was charged with the task of seeking an act of parliament to control "the regulation of passage across the River Tay, between the counties of Fife and Forfar", under the management of the Justices of the Peace for the two counties as trustees.

The document was laid before parliament and received the Royal assent in June 1819. A management committee was appointed and, on Wednesday 28th July 1819, four weeks after the passing of the act, the first general meeting of the trustees was held in the Dundee Town Hall. With a constitution drawn up, operations of the Tay Ferry Trust could commence. The trust, by powers conferred in the Act of Parliament, were entitled to claim all rights to property and the jurisdiction of ferry operations from Ferry-Port-on-Craig (Tayport) to Balmerino on the Fife side, and Invergowrie to Broughty Ferry Castle on the Forfar (Dundee) side of the river. The trustees also appointed a superintendent at a salary of £80 per annum, with a free house. His first job was to fix regular sailing times for yawls and pinnaces, and have them manned by hand picked crews. All boats were purchased, at valuation, with compensation being paid to the ferrymen for loss of employment. Thus, the number of boats was reduced from almost thirty, to eight in all.

This had been swiftly done, and the task was completed by August 1819. On the 7th September 1819, the redundant boats and pinnaces were sold by public roup in the Town Hall at Dundee. For the first time in the history of the ferry service, a regulated passage from Dundee to Fife had been established.

Steam Age

By 1820, with a rapid increase in traffic, the trustees steered towards the use of steamboats, at a time when there were only fourteen such boats registered in the whole of Scotland. A twin-hulled boat, the 'Union', was built at Perth by Jas. Brown, and fitted with steam engines built by Messrs. Carmichael of West Foundry in Dundee. The 'Union' was placed in service between Dundee and Woodhaven/Newport-on-Tay, in 1821. A second similar twin-hulled boat, 'George IV', was ordered and joined the 'Union' in 1823, by which time Carmichaels had designed a reversible motion steam engine, to allow the boat to be propelled ahead or astern. Each of the steamboats were manned by a captain, an engineer, a fireman and five seamen.

After a year of alternately sailing from Dundee to Woodhaven or Newport-on-Tay, it was decided to abandon the Woodhaven passage and to concentrate on the Newport-on-Tay crossing. This amendment had to be written into a new act and placed before parliament, since the original act clearly made a provision for the two crossings.

With good management and modern steam boats, the traffic and revenue steadily increased. On the 10th September 1823, the foundation stone for a new pier at Newport was laid. The old pier at Woodhaven was abandoned and Mr Stewart of St. Fort, its owner, was compensated for the loss of revenue.

The Union

The 'Union' and 'George IV' continued to share the passage until the 'Tayfield' replaced the 'Union' in 1836. The 'Princess Royal' was built in 1840, being constructed, like her predecessors, on the twin hull principle. The 'Princess Royal' was built by Peter Borrie at Broughty Ferry, and was launched on Friday 11th December 1840. This vessel was, at the time, the largest paddle steamer in the fleet with 3604 square feet of deck space, and being propelled by two steam engines of 80 hp each.

The 'George IV' was sold by public auction leaving the 'Tayfield' and 'Princess Royal' to ply the Tay until 1853, when the 'Newport (I)' was built. The name 'Newport' was the only one which would be repeated when naming the ferryboats. The 'Newport (I)' was the last wooden vessel to be built for the Tay Ferries, and served only until 1861, when she was withdrawn from service.

Ownership Changes

Although revenue had been increasing over the years, the fare structure was now not covering the working expenses. In 1835, the trustees, in an effort to balance the books, leased the ferries at £2000 per annum, on five year agreements to tacksmen. It was the responsibility of the lease holders to maintain the ferries in good working order. Under this management arrangement, the traffic steadily increased.

The ferries ultimately fell into the hands of the government, who handed them over to the Scottish Central Railway Company to manage. Having not much more success than their predecessors, the directors, in 1854, adopted a new timetable with regard to Sunday sailings. These additional trips proved very popular with the travelling public, but were at a great hardship to the crews and the men employed on the piers, with many of the old hands resigning. The public were alarmed by this action as, due to the requirement for a thorough understanding of the sandbanks and the currents, it was not considered that the navigation of ferries should be left to inexperienced men.

Aware of public concern, on the 24th September 1854, at a public meeting held at Newport-on-Tay, the directors of the Railway Company Owners reduced the Sunday sailings to the number only required by the Act of Parliament. These concessions were too late to prevent the retirement of the captain and mate, with their resignations already confirmed. On Saturday 28th October, at a meeting in the British Hotel Dundee, Mr. John Berry of Tayfield presented a watch and purse of ten sovereigns to the captain, and a purse of ten sovereigns to the mate, for their “esteemed service”.

On the 14th December 1858, the ‘Fifeshire’ arrived at Dundee, having been built by R. Napier & Sons at Glasgow. She was followed into service in 1861 by the ‘Forfarshire’, built by Messrs. Gourlay Bros. of Dundee. Both of these new ferries were, at the time, of modern build and were more powerful and commodious than any of their predecessors.

In November 1873, the operations of the Tay Ferries was bought by the Dundee Harbour Trustees, for the sum of £20,000. Under the new ownership, great improvements were carried out at the piers on both sides of the river. Between 1878-1884, new offices and waiting rooms were constructed, along with a new house adjoining Craig Pier, for the superintendent of ferries.

On the 28th June 1875 another new steamer, the ‘Dundee’, of similar size to the ‘Fifeshire’, entered service having been built by W. Simons & Co. at Renfrew. The ‘Forfarshire’ was used as the spare steamer and was subsequently sold in 1893 to The North British Railway Company, for their service at Queensferry on the River Forth.
The effects of the bridging of the river with a railway bridge was now taking business away, and the ferry finances were again in a poor state. In 1899, fares were fixed at a new level which was held until July 1915, with only small increases in fares being made during and after the First World War.

Picture: University of St. Andrews

With the motor lorry now established, more vehicular traffic was using the ferry crossing. This brought about the need for larger capacity boats to take advantage of this new demand, and revive the fortunes of the ferry service.

In 1910 a new paddle steamer, the 'Newport (II)', was built at Dundee by The Caledon Shipbuilding and Engineering Company, and entered service on 15th September of that year.

The 'Dundee' was sold to The Tay Steamboat Company, in 1917, for pleasure sailing on the river, for the sum of £3,000. She was later sold to The North British Railway Company for their Queensferry (River Forth) service, and was finally scrapped in 1952.

Opposite:

P.S. Dundee at Craig Pier. P.S. Dundee carried 10 cars and was built at a cost of £8970, by William Simons & Company of Renfrew

Picture: The Courier, Dundee

Greater Demand

With the introduction of bigger motor lorries and buses, greater capacity boats and improved facilities at both piers was, by the 1920's, generally recognised. At Newport, the west side of the slipway was the most convenient for landing at all states of the tide, but was found to be too narrow for the larger vehicles manoeuvring off and on the ferries. The decision was therefore taken, to double the width of the pier and also to extend it to suit loading at low water.

The 'Sir William High', built by The Caledon Company at Dundee, entered service on 21st October 1924, in place of the 'Fifeshire' which was laid aside as the spare steamer and eventually sold in April 1929 for £595, to a firm of breakers in Alloa.

With the ever increasing traffic, it was now considered that the hourly service was not coping with the demand by motor traffic and, on the 1st May 1929, two steamers began a half-hourly service (which lasted until the end of the ferries in 1966, when the Tay Road Bridge was opened).

It was in 1929 that the largest steamer built for the Tay ferry crossing, the P.S. B.L. Nairn (being 10ft longer and 2ft broader than the P.S. Sir William High) went into service, with the extra space being devoted to the foredeck in order to accommodate a greater number of vehicles. This increased capacity brought about a period of financial security once again for this fragile, yet important service.

Opposite: P.S. Sir William High - 311 gross tons

Launch of P.S. B.L. Nairn Picture: Marine Design Consultants Ltd.

'B.L. Nairn' ready for service *Picture: Marine Design Consultants Ltd.*

Picture: The Courier, Dundee

Wind of Change

With the 'B.L. Nairn' and 'Sir William High' running the regular 1/2 hourly service, and the P.S. Newport being used as the spare boat, traffic continued to increase during the 1930's. Dundee Town Council began to show concern that the ferries could not keep up with the demand, especially from commercial vehicles which were steadily increasing in size and number.

On the 12th December 1934, Dundee Town Council called a meeting of all local authorities between Aberdeen and Edinburgh, to consider a proposal for the construction of a road bridge between Dundee and the Fife shore of the River Tay. Although the seed was planted, the realisation of a road bridge was not to be for another thirty years.

In May 1937, it was decided that a larger boat was required to supply the increasing demand for passage between Dundee and Newport. It was felt that the 'B.L. Nairn' was as large a vessel of the paddle steamer design as could be operated, due to the operational difficulties presented by low water at each side of the river and the number of vehicles required to be carried, yet it still did not offer the required deck space now in demand.

Opposite: A busy 'B.L. Nairn'

Picture: Mrs L. Brown

At a meeting in May 1937, the Boards attention was drawn to the Voith-Schneider form of propulsion. A vessel with this propulsion system was already operating on Lake Constance, between Germany and Switzerland. A committee was formed which left for Europe to view at first hand the operating advantages claimed by this form of propulsion. With so many changes and the demands on the service, Mr. Wm. Johnston, superintendent of Ferries, whose health had been failing, decided to resign after thirty-eight years of service.

On the 19th July 1937, Captain L.J.D. Whitelaw was appointed superintendent of the Tay Ferries, at a salary of £350 per annum, with a free house, coal, and light. It would now be Captain Whitelaw's responsibility to see through the changes now under consideration, to ensure that the ferries would cope with the ever increasing demand.

At a meeting of the Dundee Harbour Trustees, on the 21st February 1938, the tenders for the building of a new ferry were considered. The trust decided to accept the offer of Messrs. Fleming and Ferguson of Paisley, with delivery agreed for the latter part of 1939.

In March 1938, Voith-Schneider offered the use of model propellers for tank testing, at their facilities in Hamburg, Germany. These tests revealed that if the propellers were placed fore and aft, a 50% increase in power would be required. It was therefore recommended that both propellers be placed at the rear. This was accepted by the board and the order for the propellers was placed with Voith-Schneider on the 6th October 1938, with a delivery time of seven months.

Opposite: Captain L.J.D. Whitelaw in his office at Craig Pier

Picture: J.A.S Watson, Selkirk

On the 22nd June 1939, the Harbour Trustees announced that the launch date, given by Messrs. Fleming and Ferguson, would be 4th July, 1939. As a departure from other names used for the ferry boats, this new vessel would be named M.V. Abercraig, in line with other Harbour Trust owned vessels also called 'Craig'. Mrs Henry Campbell Scarlett, daughter of the Convenor of The Tay Ferries Committee, was asked to launch and name the new ferry.

By the end of October 1939, the 'Abercraig' had arrived safely at Dundee. Two crews were trained in her operation, and after being dry-docked for a pre-service examination, she was ready for service, and started on passage in November of that year.

Opposite:

'Abercraig' arriving at Dundee - gross tonnage 455 tons. Capable of carrying 1200 passengers and 45 vehicles

Picture:The Courier, Dundee

M.V. Abercraig

'Abercraig' was only the second vessel, operating on ferry service in British waters, to be equipped with Voith-Schneider propulsion. The other vessel, 'Vecta', was built for the Cowes-Southampton service and, at the time, was the highest powered merchant ship to be so equipped. The 'Abercraig' was the first vessel to utilise horizontal propelling machinery in conjunction with Voith-Schneider propellers.

M.V. Abercraig was equipped with two eight cylinder Blackstone-Brush diesel units, designed to each develop 450 b.h.p. at 430 r.p.m., being constructed at Paisley in 1939 by Fleming and Ferguson Ltd. The design was such that large motor vehicles up to 16 ton could be carried and, at that time, up to 1200 passengers. In order to handle large lorries and buses, two electrically driven turntables of 25 feet in diameter were arranged opposite the gangways on the main deck, a complete turn taking just 60 seconds to complete. The gangways were self-adjusting, since the landings on both sides of the river were of the slipway design.

The choice of Voith-Schneider propellers had been specified for their exceptional manoeuvring abilities, allowing the vessel to stop in its own length and to literally move sideways under her own power. The propellers could also enable the vessel to be steered when she was neither proceeding ahead or astern, all in the shallow waters at the piers. The M.V. Abercraig's hull was designed to take into account strong tides, high winds and the obstruction of the course by sand banks and rocks.

Opposite: Manoeuvring using Voith-Schneider controls

Picture:The Courier, Dundee

Troubled Times

By 1941, with M.V. Abercraig and P.S. B.L. Nairn conducting the 1/2 hourly service, and the P.S. Sir William High as the spare vessel, the P.S. Newport was unused and it was decided to accept an offer by Messrs. Thomas McLaren of Glasgow, on behalf of Metal Industries Ltd., for the unused ferry. In June of that year, the 'Newport (II)' left the River Tay for the last time.

During the war years, M.V. Abercraig was to be out of service on many occasions. Although this was mainly due to propeller problems, she was removed on one occasion due to a seized piston in the port engine, which had broken up and caused grooving on the cylinder lining. At this time, mud deposits were also found in the cooling system. It was discovered that this was a drawback of the sea-water cooling system being employed and consequently, it was decided to have a fresh water tank fitted for the supply of coolant to the two engines, at a cost of £386.

With Britain now at war with Germany, the continued propeller problems could not be referred back to Voith-Schneider. A British firm, David Brown, were asked to inspect the propeller blades and to manufacture replacements. Brown's Glasgow agent reported that, in their opinion, the problems experienced were caused by inferior metal used in the original manufacture of the blades, and that much superior metal would be employed by Browns when producing the replacements.

Opposite: Turntable on foredeck of M.V. Abercraig and M.V. Scotscraig

Picture: University of St. Andrews

Although pleasure boats in various private ownerships had cruised on the Tay for many years, the Tay Ferries Trustees had not ventured into this market. Early in 1943, a decision was taken to sail a 'Fifie' (as the ferryboats operated by the Tay Ferries Trust were affectionately known) to Newburgh, during the Dundee Trade holiday week in July of that year. The 'Fifie' would dock at the pier belonging to the Tayside Floorcloth Company, allowing passengers to disembark and walk to the town of Newburgh.

In 1944 this practice was repeated, but the landing this time was to be at the pier belonging to Messrs. Bell Brothers after a dispute with the Tayside Floorcloth Company, with the P.S. B.L. Nairn sailing one trip per day at times suitable to the tides leaving from the west protection wall at Dundee.

Opposite: Abercraig at Newport-on-Tay

TAY

Specification of all 13 boats buil

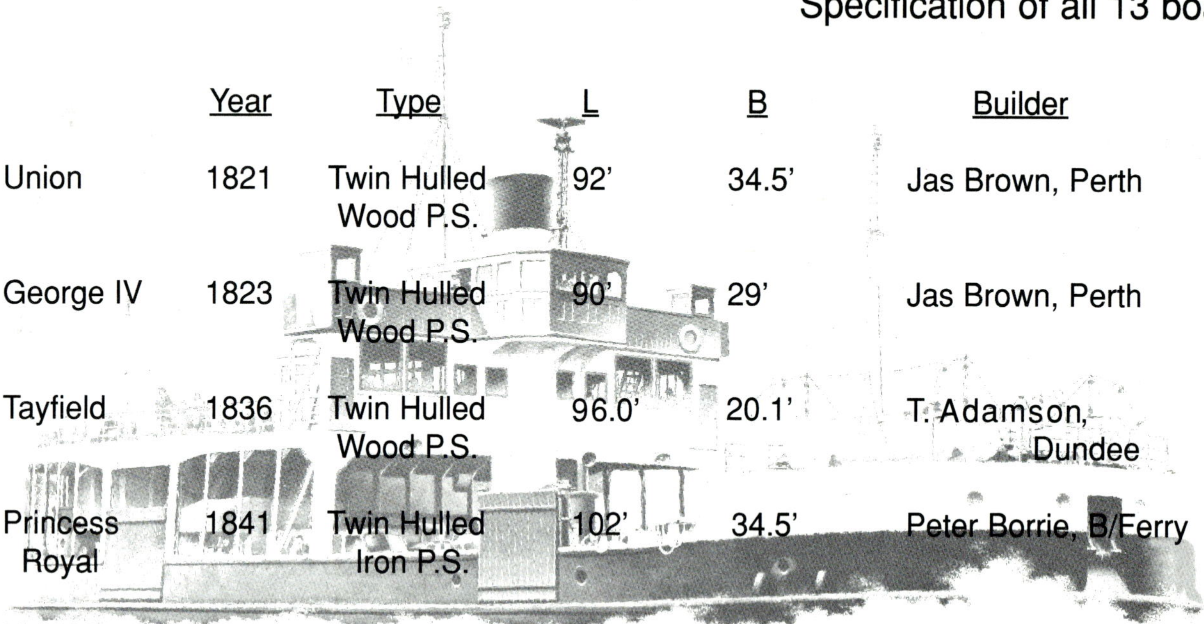

	Year	Type	L	B	Builder
Union	1821	Twin Hulled Wood P.S.	92'	34.5'	Jas Brown, Perth
George IV	1823	Twin Hulled Wood P.S.	90'	29'	Jas Brown, Perth
Tayfield	1836	Twin Hulled Wood P.S.	96.0'	20.1'	T. Adamson, Dundee
Princess Royal	1841	Twin Hulled Iron P.S.	102'	34.5'	Peter Borrie, B/Ferry
Newport (I)	1853	Wood P.S.	128.1'	23.2'	Jas Brown, Perth
Fifeshire	1858	Iron P.S.	134.4'	30.1'	R. Napier & Sons, Glasgow

FERRIES

for the Dundee -Newport passage

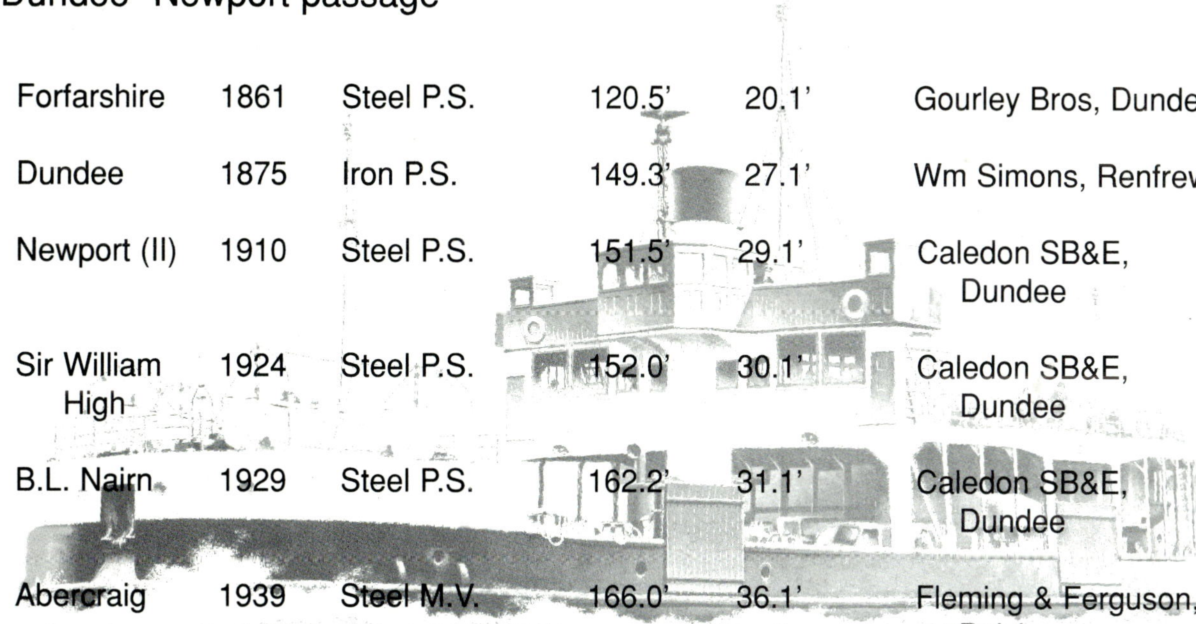

Forfarshire	1861	Steel P.S.	120.5’	20.1’	Gourley Bros, Dundee
Dundee	1875	Iron P.S.	149.3’	27.1’	Wm Simons, Renfrew
Newport (II)	1910	Steel P.S.	151.5’	29.1’	Caledon SB&E, Dundee
Sir William High	1924	Steel P.S.	152.0’	30.1’	Caledon SB&E, Dundee
B.L. Nairn	1929	Steel P.S.	162.2’	31.1’	Caledon SB&E, Dundee
Abercraig	1939	Steel M.V.	166.0’	36.1’	Fleming & Ferguson, Paisley
Scotscraig	1951	Steel M.V.	166.0’	36.1’	Caledon SB&E, Dundee

Picture: The Courier, Dundee

Opposite:
'Abercraig' undergoing a refit

Radar

Over the years, disruption of service had been experienced due to fog, resulting in the grounding of the boats on the many hazardous sand banks and rocks in the Tay Estuary. In 1947, the decision was taken to fit radar to the 'Abercraig'. This was undertaken by Messrs. Cossar Radar Ltd., at a cost of £3,000.

Within a few months, the benefits having been appreciated, Cossar Radar were asked to supply radar equipment for the 'B.L. Nairn', with the fitting of this new equipment being completed during the summer of 1948.

An accident prompted the fitting of radar into the 'Sir Willaim High'. The vessel became grounded on Fowler Rock after leaving Newport with 60 passengers and 11 vehicles on board. Broughty Ferry Lifeboat was called out, and the RNLB Mona landed the passengers at Camperdown Dock. At an inquiry into the accident, although conditions had not been perfect, the paddle steamers master was found to have made an "error of judgement". The Board of Trustees decided that with the success of the radar already fitted on 'Abercraig' and 'B.L. Nairn', it was time to fit radar to 'Sir William High', and the work was carried out early in 1949.

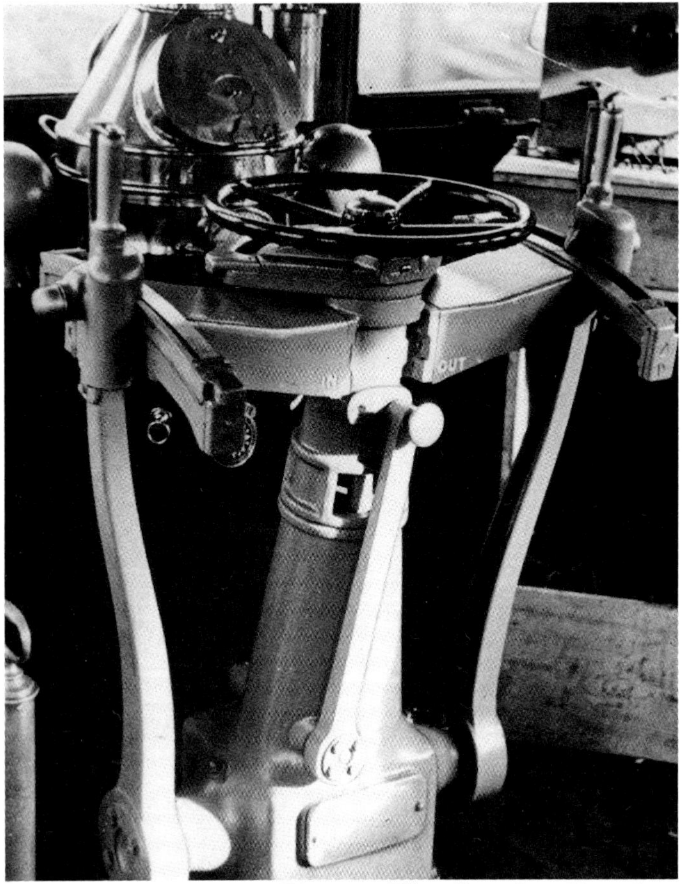

Picture: The Courier, Dundee

Opposite:
Voith-Schneider controls

Another New Ferry

With the war now over and the traffic again increasing, David Mudie, convenor of Ferries, stated at a meeting on 7th February 1946, that difficulties were experienced in dealing with heavy goods vehicles when the P.S. B.L. Nairn and P.S. Sir William High were on passage.

Initial difficulties with M.V. Abercraig having, for the time being, been overcome, it was now time to consider another new vessel to share the service with the 'Abercraig' - this new boat to have equal deck space to the 'Abercraig'. Difficulties with the Voith-Schneider propellers were noted, and the committee's opinion was that a design with less complicated propulsion would be more desirable.

In April 1946, the Caledon Shipbuilding & Engineering Company in Dundee, agreed to undertake experimental work on the design of a hull and propelling mechanism for a new vessel. As tests continued, concern was expressed over the requirement for exceptional helm control at low speeds with the shallow draft, along with the ability for a twin screw type vessel to stop in its own length as soon as the engines were reversed. This type of control was most required when approaching Craig Pier on the Dundee side of the river, and the 'Abercraig' had proved , with its Voith-Schneider system, to be very successful at this type of control.

Also under consideration at this time was the possibility of splitting the vehicular and passenger traffic. With the post-war availability of ex-government tank landing craft, it was considered that this type of vessel could prove successful at conveying vehicles across the Firth of Tay, in tandem with 'passenger only' boats.

Picture:Peter Anderson

The Caledon Shipbuilding & Engineering Co. drawing office in the 1920's. Joe Anderson, whose career spanned the years which saw the building of the 'Sir William High', the 'B.L. Nairn' and the 'Scotscraig' for the Tay Ferries Trust, is pictured at his drawing board in the centre of the picture.

M.V. Scotscraig

While still in the project stage, the ferry operators had expressed concern over the complexity and reliability of the Voith-Schneider propeller system. The Caledon Shipbuilding & Engineering Company in Dundee, were sympathetic to its clients concerns and most early designs and tests were carried out on the basis that the new ferry would be a twin screw vessel, powered by diesel-electric machinery. Since Dundee Harbour Trustees were not familiar with the capabilities of diesel-electric drive, demonstrations and tests were carried out with the co-operation of the Isle of Wight Steam Packet Company, on their diesel-electric motor vessel.

These trials confirmed that diesel-electric machinery on bridge control is greatly superior to orthodox engines when used with screw propellers, this also being so when used to drive paddles. Unfortunately, with the shallow draft required for the hull design, due to the operational difficulties on the River Tay, cavitation reduced the braking effect of the screw propellers, to the extent of introducing an "element of risk", despite the rapidity of response of the diesel-electric drive system.

Again, Voith-Schneider propellers were placed under consideration. Experience with M.V. Abercraig had proved that the manoeuvring and braking requirements could be met with the German system. Unfortunately, this system being susceptible to damage, had proved detrimental to maintaining a reliable service.

Picture:The Courier, Dundee

With no realistic alternative, the decision to build an almost sister ship to the 'Abercraig' was taken. The cost of a spare set of propellers would make economic sense if they could be applied to two ships and would therefore reduce down-time due to instant replacement, which could be carried out without dry-docking the vessel. The Scotscraig was equipped with a pair of 6SRKM English Electric diesel engines, since the diesel-electric option was no longer required with the decision to use the Voith-Schneider propeller system which did not require reversing engines.

After all the options had been considered, tenders for a new ferry boat similar to M.V. Abercraig were invited from shipbuilders. In May 1948, a bid of £152,450 by Caledon Shipbuilding & Engineering Company at Dundee was accepted, with delivery to be made in three years.

Opposite: 'Abercraig' leaving Newport-on-Tay

Pictures: Marine Design Consultants Ltd.

Increasing Workload

With the order now firmly placed for the new ferryboat and the keel already laid, in August 1950 it was decided to seek offers for the P.S. Sir William High, with delivery when the new ferryboat entered service, thus leaving the P.S. B.L. Nairn as the spare boat.

During the early part of 1951, while the 'Scotscraig' was still being fitted out, a new material, 'Wearite', became available. Exceptionally easy to clean and resistant to cigarette burns, it was decided to use this material, when available, as panelling in the saloon area of the new vessel.

The 'Scotscraig' was launched in May 1951, by Mrs. F.D.I. Buist, wife of the Tay Ferries Convenor, and entered service on Monday 3rd September 1951, with temporary panelling in her saloon awaiting delivery of the 'Wearite' material.

Also at this time, the 'Abercraig' was again having piston and liners fitted due to excessive wear, and two spare propeller units were ordered for use, as agreed, in both the ferries.

In December 1952, with the two motor vessels maintaining the 1/2 hourly service between Dundee and Newport-on-Tay, and the P.S. B.L. Nairn on standby, the P.S. Sir William High was sold for £15,000, to Messrs. Ojukwo Transport Ltd., for service in Nigeria.

Opposite: Launch of the 'Scotscraig' by Mrs F.D.I. Buist

During the night of 15th February 1953, with an exceptionally low tide, the 'Scotscraig', while berthed, rested on the bottom at Craig Pier, piercing one of her bottom plates. She was dry-docked and out of service for a month.

Apart from a period during the 1940's, cruising on the Tay had been left to private boatmen to satisfy the market. However, in celebration of the coronation of Queen Elizabeth II, Dundee Corporation asked the Tay Ferry Trustees to provide an evening cruise to Balmerino, on the Fife side of the river for 200-300 senior citizens, aboard one of the 'Fifies'. This was carried out with refreshments onboard being provided by the Dundee Corporation, and the cost of the ferry at the expense of the Harbour Trust.

Early in 1953, with bottlenecks being experienced at both sides of the river, especially when P.S. B.L. Nairn was on passage, the secretary of the Scottish Board of Trade was prompted to write to the Tay Ferries Trustees concerning the delays. He enquired as to whether another large ferry-boat may be required and suggested that public funds may be available to assist in the purchase of another vessel. This, however, was never to be.

In 1954, 'Abercraig' was again suffering propeller blade problems due to corrosion. Voith-Schneider, in Germany, were asked to supply new blades of similar material to those fitted to the newer 'Scotscraig' propeller units, which were less troublesome. In 1956, a building belonging to Messrs. George McLean, adjoining Craig Pier was purchased and used as a workshop to service and repair the Voith-Schneider units which could be fitted in a short time, with the availability of the spare parts now stocked, and fully serviced with no need to dry-dock the vessels, this being the most attractive feature of the system.

With more discussion on how to reduce bottlenecks and increase traffic flow on the north-east highway, of which the Tay Ferry passage was part, more calls for consideration of a road bridge over the river were voiced. Towards the end of 1955, the trustees decided to seek a meeting with Dundee Corporation to discuss the loan debts incurred in the operation of the ferry service in the event of the road bridge being built and the ferry service being made redundant.

M.V. Scotscraig

Picture: The Courier, Dundee

Pictures: The Courier, Dundee

Skipper 'Sandy' Mands watches as Kerrs famous miniature bus boards the 'Scotscraig'. Mr Kerr was charged the motor cycle rate for the journey since the bus was not fitted with reverse gear

Accidents

With the mixture of people, vehicles, short turnaround times and navigating hazards in the Firth of Tay, great care had to be taken at all times to ensure safe passage of travellers using this necessary link between North Fife and Dundee.

The organised sailings and safety measures had after all been instigated after the accident in 1815, when 18 people were drowned. Although accidents were rare due to the dedication to duty, and concern for the safety of the passengers by the ferry crews, on 31st October 1954, while on passage from Newport to Dundee, the vessel 'Abercraig' had to put about when a man fell overboard. With great skill the ferryboat was brought alongside the passenger in the water. Crewman Frederick Rae dived overboard and supported the man until both could be hauled on board the 'Abercraig'. Frederick Rae was given a 'monetary award', which was usual in these circumstances, for his courageous act.

Again on 17th May 1955, the devotion to duty of another crewman was to save a life. While berthed at Newport, an elderly passenger fell between M.V. Abercraig and the pier. Deck hand James Mann immediately jumped into the water and assisted the distressed passenger who was taken ashore further along the slipway. For his action, which undoubtedly saved the passengers life, James Mann was also given the standard 'monetary consideration'.

At 8.20am, on 26th February 1960, the P.S. B.L. Nairn and the M.V. Abercraig were in collision, in dense fog midway between the Middle Bank buoy and the Drummond Rock buoy. The 'B.L. Nairn' had left Newport on the 8.00am passage with 'Abercraig' running behind schedule due to loading difficulties at Craig Pier. With both radar screens operational, the two boats slowed as they approached each other. As they manoeuvred closer, and with loss of picture due the narrow radar beam at this range, the master of the 'B.L. Nairn' brought his vessel to a stop to allow the 'Abercraig' to pass. The two vessels came together with a slight impact on the sponson belting. With no further damage to the vessels or injury to any of the passengers, the two boats continued their journeys. At the inquiry into the collision, it was decided that the masters of both vessels had taken the necessary steps to avoid a serious accident and that "no action be taken" against the skippers.

In June 1962, several cars were damaged when children playing on the car deck of 'Abercraig' interfered with the turntable cut-out switch which set the turntable in motion. The claim for the damaged cars was handled by the insurance company and the master was reprimanded for allowing the control mechanism to be left in the 'on' position.

The 'Abercraig' appeared to take umbrage at the building of the Tay Road Bridge, when on 8th July 1964, with a fierce westerly gale blowing, her engines stalled and she collided with the temporary bridge used to carry materials out over the river to the builders. Workmen on the bridge assisted the passengers to safety along the temporary bridge to the Dundee shore. Bridge builders Duncan Logan subsequently made a claim for the structural damage caused by the collision.

Abercraig at Newport-on-Tay *Picture: University of St. Andrews*

Journeys End

With the ever increasing strain on men and vessels, the P.S. B.L. Nairn, was beginning to feel her age. Having entered service in 1929, she had proved a capable and reliable boat. In February 1957, the main drive shaft on the starboard paddle fractured. At a later survey, close examination showed that extensive re-plating of the hull would have to be carried out immediately with the probability of more in the future if she was to be retained in a fit condition, as the standby vessel.

The late 1950's saw yet another senior technician come to Dundee from Voith-Schneider in Germany, with both 'Abercraig' and 'Scotscraig' experiencing mechanical problems. On arriving for duty on 15th June 1960, the crew found M.V. Scotscraig's hull to be holed with water entering the engine-room while she had been laid up over night. Divers were sent down to examine the bed of the river at Craig Pier, since it was thought she may have settled on a sharp object at low tide. With no such object found, concern that electrolytic action could be causing pitting of the hull plates prompted the trustees to contact Messrs. Duff & Partners Ltd. of Hull to seek their advice. With a plate welded over the area, the whole of the hull was treated, on Duff's advice, with special paints. 'Abercraigs' hull was, by 1961, also showing its age and five new plates were fitted during its annual survey.

Although the 'Abercraig' had been plagued with engine troubles and was undergoing a £17,000 engine overhaul, over the years the 'Scotscraig' had been relatively trouble free in this area. It was therefore a cause for concern when the starboard engine suffered a breakdown in June 1961.

Picture: The Courier, Dundee

The Tay Road Bridge

On the 18th August 1966, the Tay Road Bridge was officially opened by H.M. The Queen Mother, with the hope that delays and bottlenecks would be a thing of the past. With a steady increase in the volume of traffic using the bridge at peak times, especially in the mornings, tail-backs on the northbound carriageway became frequent. This eased however, when it was decided to remove the toll booths from the northbound carriageway and charge double toll for vehicles travelling south.

So with 140,000 tons of concrete and box girders utilising 8150 tons of steel, and the hard and often dangerous work of 420 men over the 3 years it took to build, with a cost of £3,823,153.17.9d out of an overall project cost of £6 million, the much talked about, 11/2 mile long, 60ft wide bridge saw, with mixed feelings, the demise of the proud and highly respected ferry service.

The ferryboats were retired to Victoria Dock shortly after the opening of the bridge, to await their fate or future.

Opposite: Road Bridge nears completion as 'Scotscraig' heads for Dundee

Picture: The Courier, Dundee

Epilogue

The Ferry crewmen who were due to retire left for home, with the younger men, who had been lucky enough to secure a job with the Road Bridge Board, starting another phase of their careers on the river as tollmen and inspectors.

The three ferryboats, now in Victoria Dock, were advertised for sale. The P.S. B.L. Nairn was first to go, sold for scrap and towed on her last journey to Blyth on the Northeast coast of England. The P.S. B.L. Nairn could be considered the most hard worked and most reliable vessel ever built for passage on the Tay Ferry route from Dundee to Newport. She had served all her long life on the River Tay with little down time due to breakdown, having been built in 1929, by the Caledon Shipbuilding and Engineering Company at Dundee. Her 37 years of service had gained her the respect of the crews, although she was a bit hard to handle in high winds, and did have problems loading vehicles at low tide, especially at Craig Pier. It was a sad day for all concerned when the last paddle steamer built for the Tay Ferries left on her final journey.

Opposite: 'B.L. Nairn' awaits being towed away for scrap

Picture: The Courier, Dundee

Although there were several interested parties considering the two 'Craigs', the Irish Tourist Board seemed to be the most likely buyers. Representatives visited Dundee to view the boats and to study the costs of running the vessels with the assistance of the Harbour Trustees. The board were in no hurry to sell, with a top price for the two vessels their main concern. The months passed, and the boats were advertised again and this time a buyer was found. The two 'Craigs', and the considerable arsenal of spares, were sold for a total of £15,000 to a Portsmouth Company.

Craig Pier at Dundee was sold to the town Corporation for £48,000, with the surrounding ground being used for road widening. Many years later, the old slipway was transformed into a luxury custom built berth for another Tay built boat - Captain Scott's 'Discovery', managed by the Dundee Heritage Trust as Dundee's flagship tourist attraction.

The two 'Craigs' were sold on and travelled to a new life in the warm waters around Malta, experiencing a variety of uses. On one occasion, the new owner, Mr. Salv Bezzina, took a party of friends to the neighbouring island of Comino (the smallest of the Maltese Islands), transforming the vessel into a floating function suite, with the car decks decorated with coloured lights and tables with umbrellas filling the area once occupied by the cars of Fife and Tayside.

This was not, however, the first time that someone had thought of using the ferries as a floating

Opposite: The two 'Craigs' in retirement

'B.L. Nairn' leaves on last journey Picture: The Courier, Dundee

'Scotscraig' leaves for Plymouth Picture: The Courier, Dundee

Opposite: The 'Craigs' tied up at Plymouth
Picture: John Sharp, Sidcup, Kent

The 'Craigs' arrive in Malta

dancehall. Mr. J. M. Wallace, owner of Dundee's JM Ballroom, had applied to hire a Tay Ferryboat on Sunday 16th July 1961, this being during Dundee Civic week. The cruise would have taken patrons up river with dancing to live music and light refreshments being served. The trustees showed a lack of enterprise and declined, citing "practical difficulties" for their refusal to make a 'Fifie' available for pleasure cruising.

When the 'Scotscraig' was at the end of its life, it was used on the film set of the Walt Disney movie 'Popeye', and was finally laid to rest on the sea bed at Anchor Bay, off the coast of Malta.

That left the M.V. Abercraig the sole survivor of the Tay Ferry fleet. Ironically, the most unreliable and accident prone of all the 'Fifies', she had at some stage been converted to front load with a ramp from the 'Scotscraig' replacing the bow section and with the midship section removed, her wheelhouse was moved aft giving her the appearance of a landing craft rather than the most advanced ferry she once was when she entered service in 1939.

Sightings of 'Abercraig' in Marsa Harbour by holidaying Taysiders, despite her ever increasing derelict state, were frequent. After a sighting in November 1994, a public meeting was held in Dundee and a steering committee was formed with the hope of saving the last 'Fifie' still afloat - well almost afloat! Meetings were held between the committee and all local funding bodies and authorities, to try to raise funds to cover the costs of a full survey to ascertain whether the 'Abercraig' could safely be returned to the River Tay.

'Abercraig' enjoying the warm waters of Malta

'Abercraig' in 1994 at Marsa, Malta

The end for 'Abercraig'

Picture: Stewart Smith

Picture: Stewart Smith

Although Tayside Regional Council showed the most interest, they would not go alone and insisted that other backers had to be found. The Abercraig Action Committee's secretary, Mr. Bradley Borland, had been busy gathering application forms from national bodies in the hope that funds could be made available to save this important part of Dundee's heritage. With pressure from the Marsa Port Authority for Mr. Bezzina to clean up his area of the harbour, time had run out for the last surviving Tay Ferry when, in June 1995, Dundee's local newspaper published a picture of the 'Abercraig' being taken apart for scrap.

With the last 'Fifie' gone, nothing was left except the memories of those who remember, with affection, the proud boats and their crews, plying the Firth of Tay in all weather conditions.

However, when a new work/rescue boat was built in 1995 for the Tay Road Bridge Joint Board, General Manager John Crerar accepted a suggestion from one of his members of staff to name the boat 'Fifie', as a lasting tribute to the Ferry Boats.

Bibliography

Body, G, *British Paddle Steamers,* David & Charles Publishers Ltd, Newton Abbot, 1971.

Brodie, I, *Steamers of the Forth,* David & Charles Publishers Ltd, Newton Abbot, 1976.

Brown, A, *Lymington - The Sound of Success,* Alan T. Condie Publications, Nuneaton, 1988.

Duckworth, C & Langmuir, G, *Railway and Other Steamers,* T. Stephenson & Sons Ltd, Prescot, 1968.

Robins, N, *The Evolution of the British Ferry,* Ferry Publications.

Weir, M, *Ferries in Scotland,* John Donald Publishers Ltd, Edinburgh, 1988.

Archive Document, *History of The Tay Ferries 1713-1930,* Dundee City Archives.